Hand to Hand Combat

Francois D'Eliscu
Lt. Colonel, Infantry

10 9 8 7 6 5 4
ISBN 0-87947-414-9
Printed in the United States of America

Desert Publications
El Dorado, AR 71731-1751 U. S. A.

INTRODUCTION

The cold blooded will to win with disregard to all rules is a challenge to our American soldiers. The realization that a man who is mentally alert, more clever, and more agile than his opponent will not be defeated, has been proven, and now appreciated by our men. They fear no enemy. Hand-to-Hand Combat provides the soldier with a new brand of physical confidence.

Our enemies have been engaged for years in a systematic program of killing. They have been taught and trained. We have discussed, studied, and improved, so that our techniques are more effective and our methods respected.

Commanders question the ultimate value of the soldier who possesses a high degree of technical skill, but is unable to stand up under the great strain of modern combat because he is not so well developed as his opponents.

The art and techniques involved in the use of the hands and body in Hand-to-Hand Combat requires a knowledge of balance, quick turning, pivoting, transferring of weight, change of pace, hitting the ground and immediate action.

Sensitive or nerve points, vulnerable spots, lever angles and all areas where serious injury is possible and practical should be studied with care.

Our attitude and personal feelings in regard to sportsmanship and fair play must be changed. Strangling and killing are remote from our American teachings, but not to our enemies. We must be prepared to kill — kill or be killed — has been found necessary.

Practice for speed and perfection. Be cautious. Do not take advantage of your partner in practice. Save your own personal techniques for the enemy!

It should be borne in mind that any sort of weapon is better than no weapon at all. Sticks, stones, a cartridge belt used as a whip, the helmet as a bludgeon, trench tool, rope or wire, knife, bayonet, machete' and many others—all become effective weapons.

Mastery of Hand-to-Hand cannot be acquired by observation alone. Personal and persistent participation under expert leadership is necessary. The beginner must observe, imitate and understand the basic movements and anatomical and physiological implications involved in gripping, squeezing, pushing, twisting, bending, tripping, throwing, pulling, etc., in order to take advantage of an opponent by surprise and an effective attack.

Fight with your bare hands only when you have no alternative. When that time occurs move fast, hit hard, apply pressure and make certain the job is completed.

☆

VULNERABLE POINTS

The techniques used in blinding, fracturing, strangling, dislocating or paralyzing an opponent, depend entirely upon the type of hold that is used, and the amount of pressure exerted. Study the chart on the right very carefully. Note the strong points to use pressure. Practice until perfection has been reached.

a) **Eye.** Never practice any kind of pressure on the eyes. This form of fighting should be used only if it offers the last means of getting out of a tight strangle.

b) **Ear.** Quick hacking of ears with hands cupped for suction causes blackout and extreme stinging pain.

c) **Nose.** A blow on the upper part of the bridge of the nose, on the side of the nose, or a quick thrust under the nose, will cause pain, bleeding, and tears to the eyes.

d) **Hair.** Pulling hair on head by "knuckle pressure" may cause paralysis.

e) **Atlas and Axis.** Any dislocation or breakage means unconsciousness or death.

f) **Back of Neck.** Any blow directed correctly at the back of the neck with the hand or club may result in serious injury.

g) **Larynx.** Finger pressure causes coughing and strangulalation.

h) **Sternoclavicular Joint.** Finger pressure or a quick hacking blow in the hollow of the throat will cause coughing, strangulation, and, at times, death.

i) **Carotid Artery.** Continued pressure on the carotid artery causes paralysis of the area.

j) **Clavicle. Pressure on the superior portion of the clavicle** close to the chest wall will cause the person to sit down immediately. The pain is evident, causing the mouth to open.

k) **Shoulder.** Pressure and pulling cause dislocation.

l) **Ribs.** Pressure on ribs will cause ribs to crack and, at times, cause internal hemorrhages. A rib injury will make it difficult for a person to breathe or bend.

m) **Solar Plexis.** Two-finger or fist thrust will disable a person temporarily.

n) **Abdomen.** A direct punch above the umbilical area will cause nausea.

o) **Spine.** Kicking, pushing, or pressure on spine will cause serious paralysis. **Dangerous.**

p) **Elbow.** A blow on the elbow joint or pressure applied there will cause breakage, dislocation and pain.

q.) **Kidney.** Very painful area with judo methods properly applied. **Dangerous.**

r) **Hip.** Quick kicking and twisting cause dislocation.

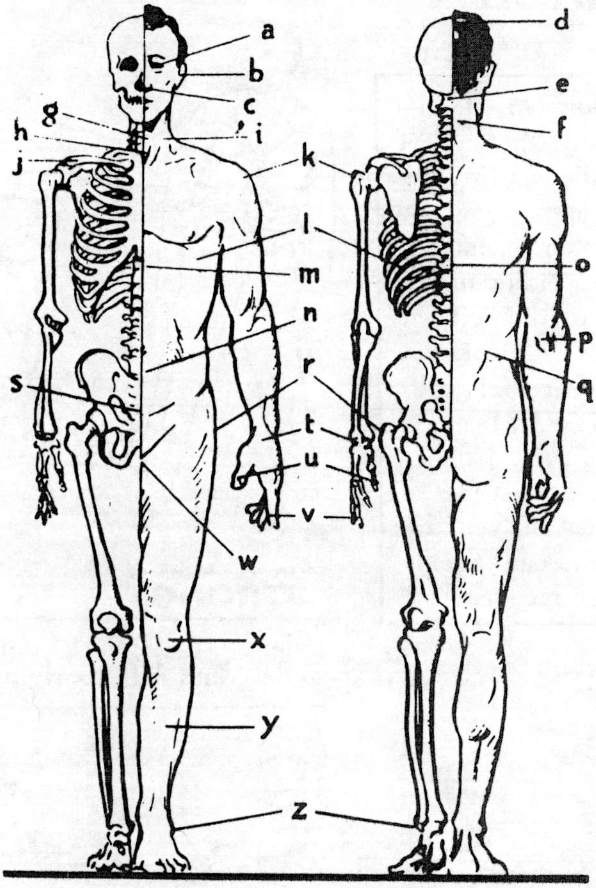

s) **Bladder.** Unless an expert, do no attempt attack. **Dangerous.**

t) **Wrist.** One of the weakest parts of the body. Very painful under any kind of pressure.

u) **Thumb.** Can be easily broken with simple twisting and applied pressure.

v) **Fingers.** Very vulnerable because easily gripped and easily twisted.

w) **Groin.** Injury is painful and **dangerous.**

x) **Knee.** Weakest part of legs because of kneecap. Easily dislocated by kicking or tackling.

y) **Shin.** A kick in the shins is painful and very effective. As opponent bends down to rub shin, follow through with quick hack on the back of the neck or knee to the face.

z) **Ankle.** Very vulnerable to kicking, twisting, or bending.

Specific Vulnerable Points of the Face. Hacking of ears, nose, eyes, above lips, teeth, over mouth, chin, side of face (mandible), under chin, in back of ears, lower portion of back of the head, on top of the head and hair.

BALANCE

Learn how to stand—never with both feet flat together or in close heel stride position. Be in a position to make a quick turn either to the right or to the left, or a complete pivot. A jump, dive or fall calls for varied balanced stances. Your balance and stance must be so strong, that if you are pushed, you can quickly counter or recover.

Balance is strong from right to left and left to right.

Balance is weak forward.

Balance is weak to the rear.

4

Balance is strong from right front to left rear and from left rear to right front.

Balance is weak to the right front

5

Balance is weak to the left rear.

6

BALANCE

STRONG

7

Balance is centered on both hips.

WEAK

8

Balance on ground is weak. Opponent can roll, kick or turn.

9

STRONG

Balance on ground is very strong. Note: Arms, legs and body locked.

LESSON **1**

ALERTNESS EXERCISES

In combat "almost right" is not good enough; you probably won't have a second chance. The exercises are a combination of calesthenics, tumbling, boxing, stunts and wrestling, which are combined with original exercises to make them interesting, and to ensure proper and effective conditioning.

10

In-fighting

11

Push and Pull Wrestling

 LESSON I CONTINUED

LESSON 1

ALERTNESS EXERCISES

12

Toe Stepping

Slap Boxing

13

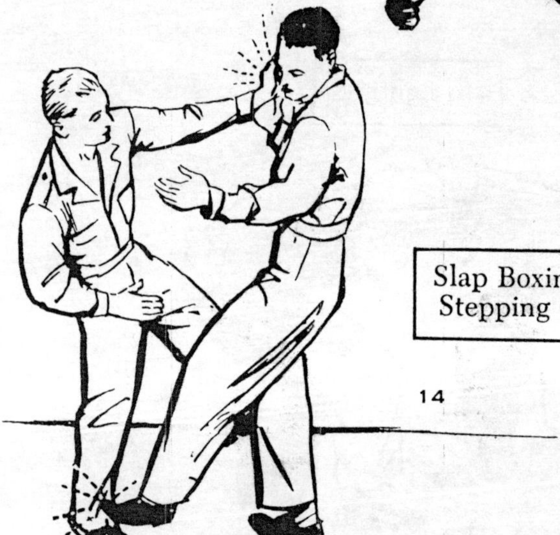

Slap Boxing and Toe Stepping Combined

14

FALLING PRACTICE

Tumbling and falling are fundamental techniques that should be acquired by every soldier. Knowing how to fall is a safety procedure, when thrown suddenly or tripped. It is also an excellent conditioner, bringing all the muscles of the body into play and developing agility and coordination to a high degree. Basic tumbling movements develop greater self-confidence. A man who knows how to handle the body in falling, is less liable to slip or lose his balance at a crucial moment because his sense of balance is well developed.

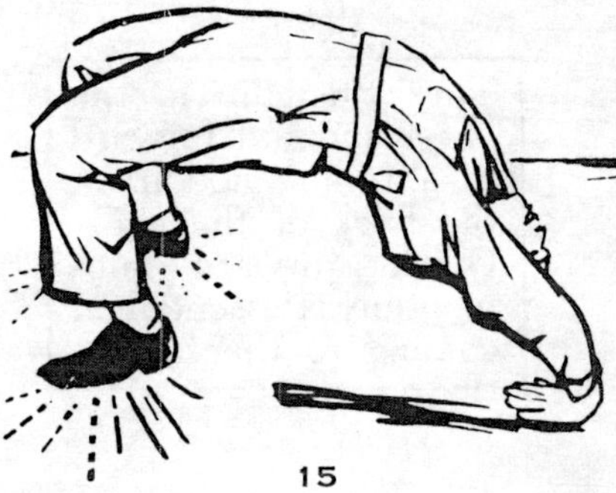

15

Feet strike first, breaking the fall. Head does not touch the ground.

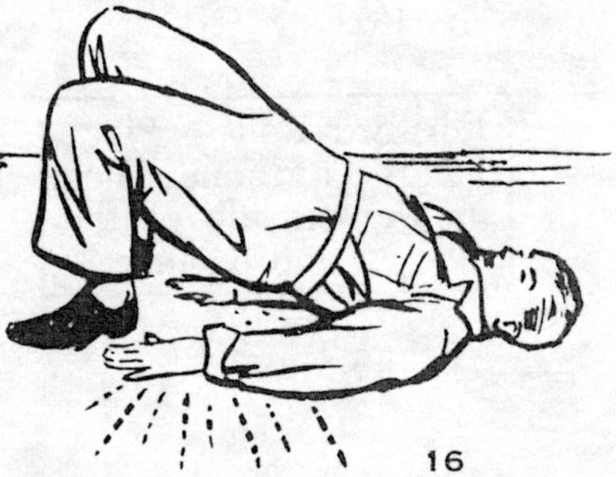

16

Arms follow immediately in a breathing action, with straight arms striking the ground before the shoulders hit.

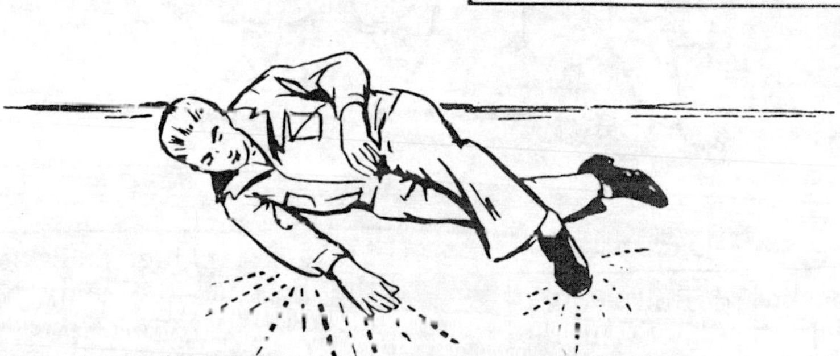

17

Side Fall—Fall is broken by a beating action of the near arm and far foot as the body strikes.

LESSON 1
FALLING PRACTICE

18

19

Forward Roll – Place hands on ground, knee on the outside of the elbows, and balance.

Tuck the chin to the chest and fall forward in a rolling action on the back of the neck. Quick forward momentum is essential in coming to the feet.

20

21

Backward Roll–Squat with the knees inside the arms.

Make a rocking chair of the back as the body is kicked forward. Backward momentum plus a vigorous thrust and push of the arms will bring recovery to the feet.

FALLING PRACTICE

22

Rolling Fall – Pull right arm between legs, fingers of both hands turned inward preparatory to rolling over right shoulder.

23

Roll from right shoulder to left buttock, crossing right foot over the left.

24

As the body rolls forward the left arm (straight) beats, bringing the body upright with the feet crossed.

25

As the body comes upright with the feet crossed, the body pivots naturally so as to face opponent in the "On Guard" position.

LESSON **1**

BLOWS AND KICKS·

Study the pictures. Practice the movements slowly for accuracy and control — then use speed and an effective follow-thru.

Heel Kick to the Solar Plexus

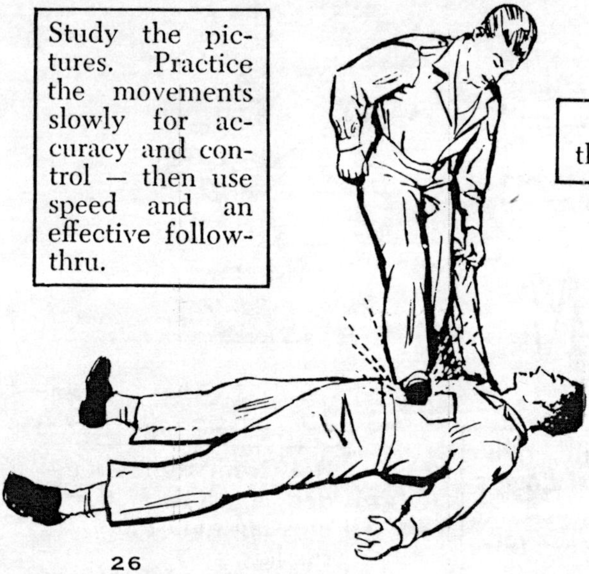

26

Toe Kick to the Neck or Head

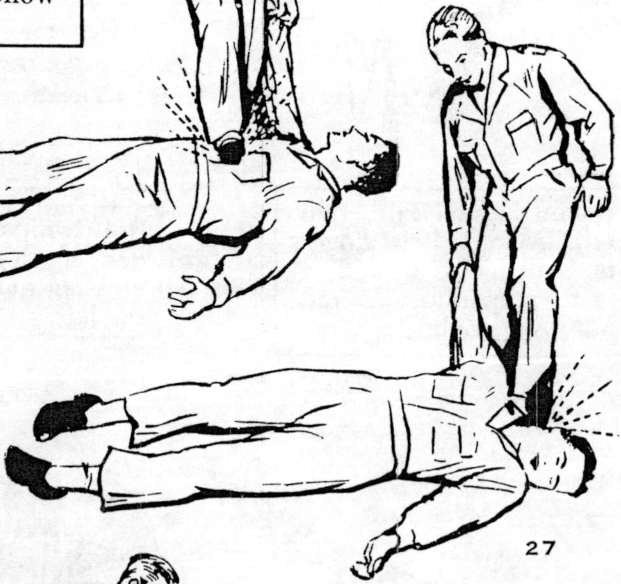

27

Heel Kick to the knee.

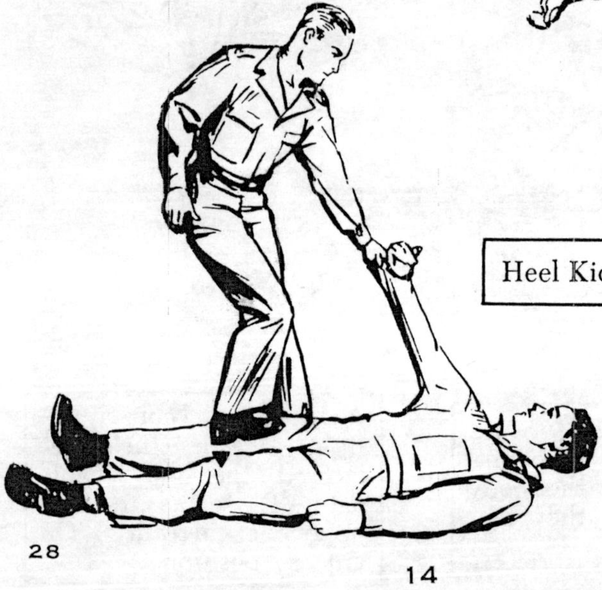

28

Kick to Floating Ribs or Kidneys

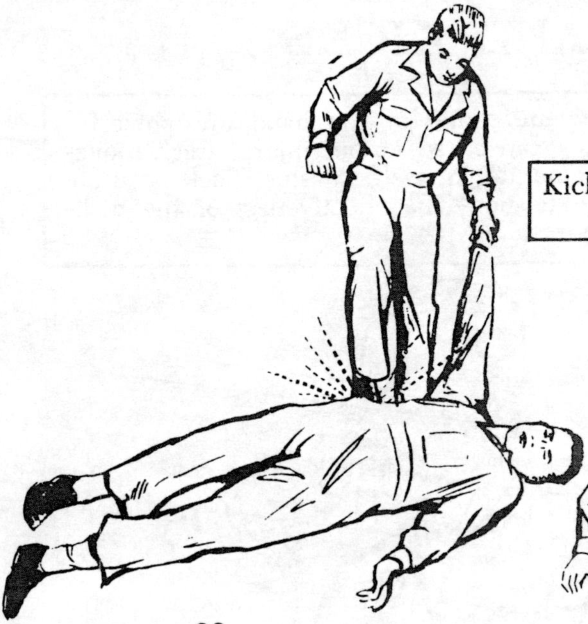

29

Heel Kick to the Center of the Back

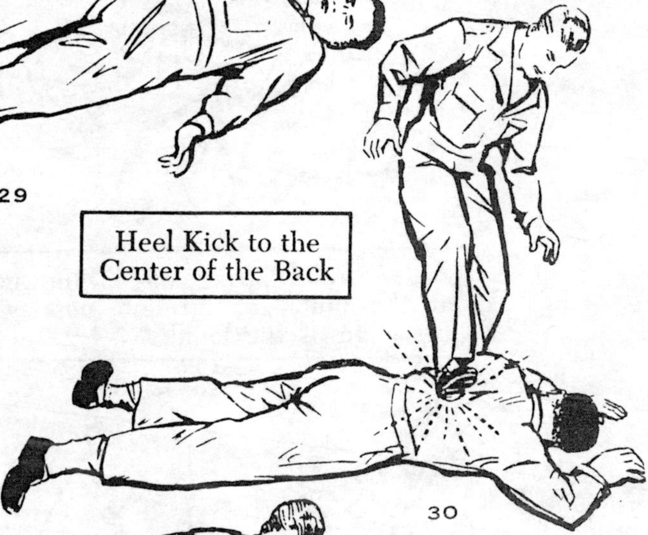

30

Heel Kick to Groin

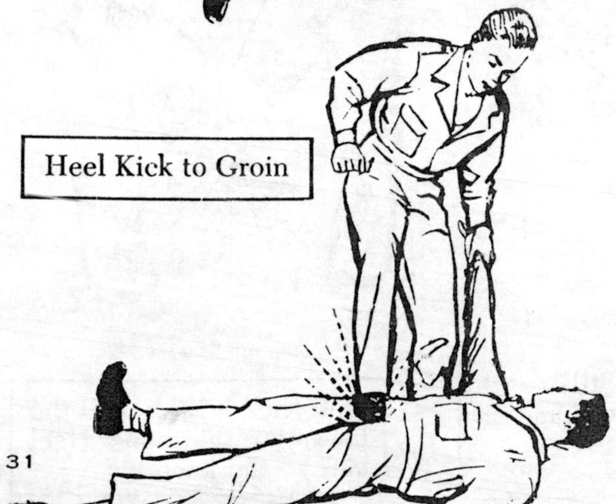

31

LESSON **1**

BLOWS AND KICKS

The fingers and edge of the hand are powerful weapons. They can cause serious injury. Each movement is a technique in itself. Master quick accurate blows to a vital spot. The effectiveness of the technique—is the result.

32

Nose Stroke—Strike under the nose at the union of lip and nose with the edge of the hand.

33

Kidney Blow—Strike with the edge of the hand in the back, over the hip bone.

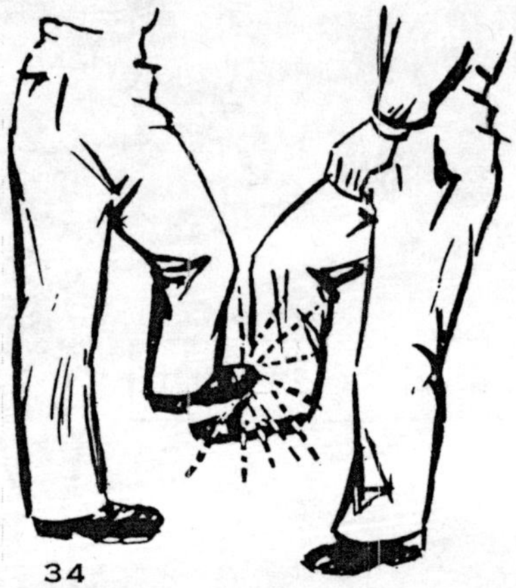

34

Shin Kick – Kick with the toe to the shin bone. Heel kick to the knee.

35

Eye Stroke – Over the nose directly between the eyes with the edge of the hand.

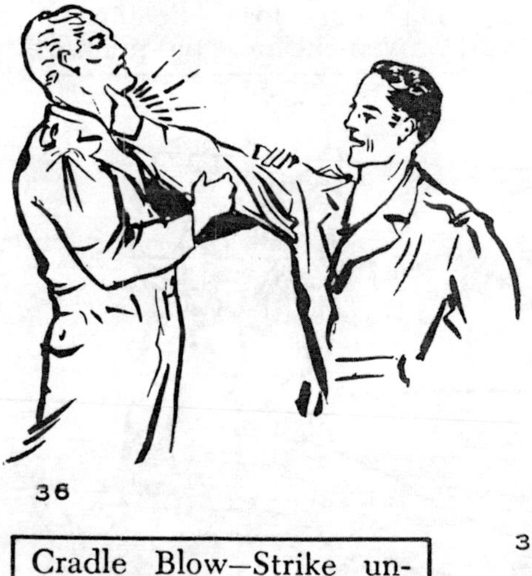

36

Cradle Blow–Strike under the throat between the "Adams Apple" and the jaw line with the open hand. Contact is made with the space between the thumb and forefinger

37

Solar Plexus Blow–Strike with the clinched fist or fingers directed at the solar plexus. Follow with knee to the Groin.

END OF LESSON 1

LESSON **2** ATTACK FROM THE FRONT

CROSS HOCK THROW

38

39

Grasp opponents left lapel with right hand and right sleeve above the elbow with left hand. Step back with the left foot.

Balance on the left foot and raise the right one preparatory to kicking or slicing his right leg. Do not lose balance. Movement must be quick.

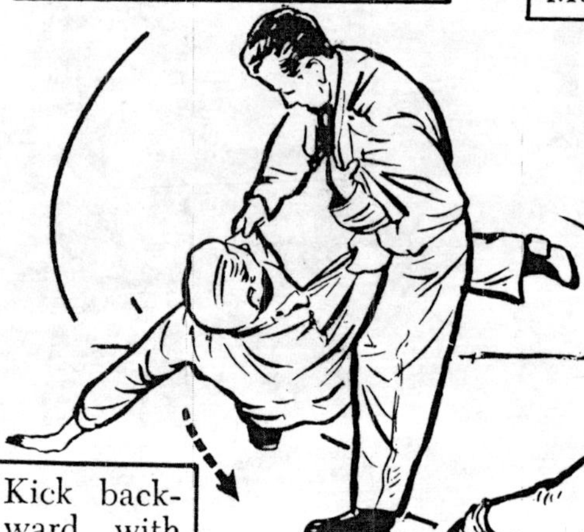

41

Kick backward with the right leg striking the back of opponent's knee. As the kick is executed the right hand pushes on the shoulder or strikes under the chin. Bring opponent to ground.—Don't lose him.

As opponent strikes the ground a kick is delivered to the back with either foot. Do not let him roll. —Keep his back to you.

STRAIGHT LEG THROW

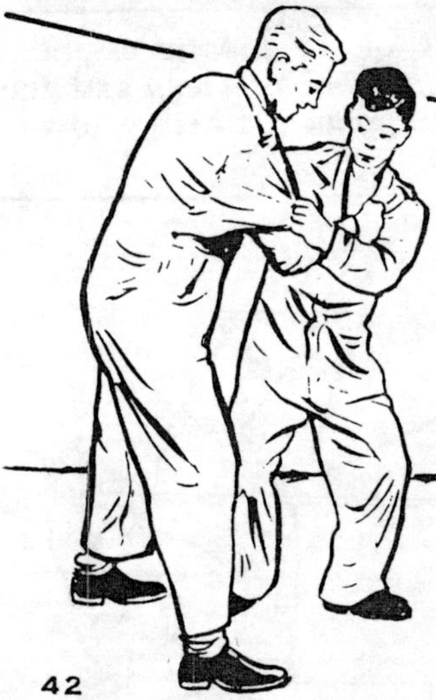

42

Grasp right lapel or sleeve. Assume a heel to heel position of the feet. This movement must be rapid and accurate.

43

Pivot to the left turning back on opponent and placing the right foot to the right of opponent's right foot, and pointing the same way, locking his knee.

44

Pull opponent forward and as close to your body as possible, throwing him to the ground.

45

Follow through immediately by dropping the right knee to the ribs and left knee on head, breaking the arm over the right thigh, and causing immobility.

LESSON **2** ATTACK FROM THE FRONT

HIP THROW

Pivoting, throwing, and the use of the hips is essential in hip throwing. The straightening of the legs and the quick pull over will bring your opponent forcibly to the ground.

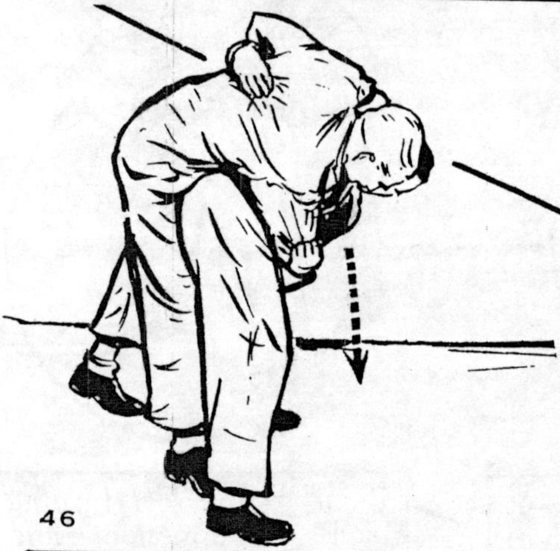

46

47

When pivoting into position the back is turned to opponent quickly and the buttocks placed against his right hip. The right hand is under his left arm and holding him. The right forearm is grasped above the elbow.

The throw is executed by straightening the legs, bowing from the waist and flipping with the hips, holding on to maintain contact.

48

As opponent falls, drop the right knee to his ribs and the left knee to the side. – Lock and break arm over left bended knee.

SHOULDER THROW

Flipping with the hips and making your opponent get off balance is necessary. Hold on to your opponent when he lands, so you can follow through.

49

Head to head position. Turn back on opponent and holding his left lapel with the right hand place the right forearm under his right armpit. Lower the hips by bending at the knees.

Throw by straightening the knees, bending at the waist, and flipping with the hips.

50

As opponent strikes the ground follow through with an edge of hand blow to the neck.

51

COUNTERS TO LOW CHARGING ATTACKS

Quick sidestepping, alertness on the part of the individual, and immediate follow-thru will counter any kind of low charging attack.

52

Drive open hand against top of opponent's head forcing his face into sharply upraised knee. Straight kick to groin or face. – Follow through with both leg jump to back.

53

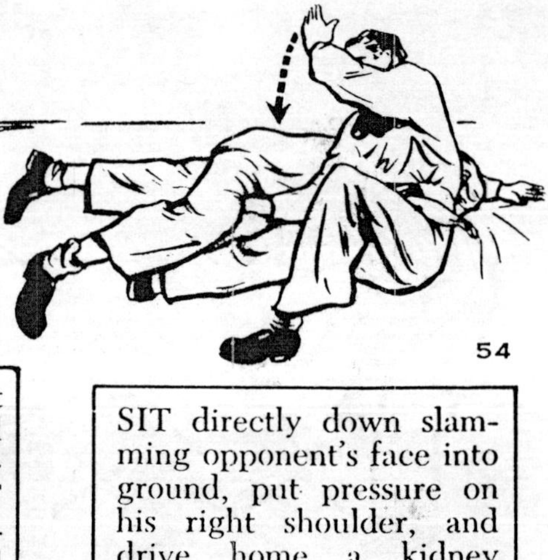

54

Grasp opponent's right elbow with left hand and place RIGHT hand inside opponent's RIGHT thigh. Positions are reversed if opponent's head is on left side.

SIT directly down slamming opponent's face into ground, put pressure on his right shoulder, and drive home a kidney blow.

55

56

Headlock: Opponent charging low — grasp head — lock right hand around neck. Place left hand on opponent's right shoulder. Right wrist grasps left wrist.

Apply pressure by locking arms tightly and raising head which is fixed and close to the body. (Dangerous.)

COUNTERS TO BEAR HUG—
OVER ARMS

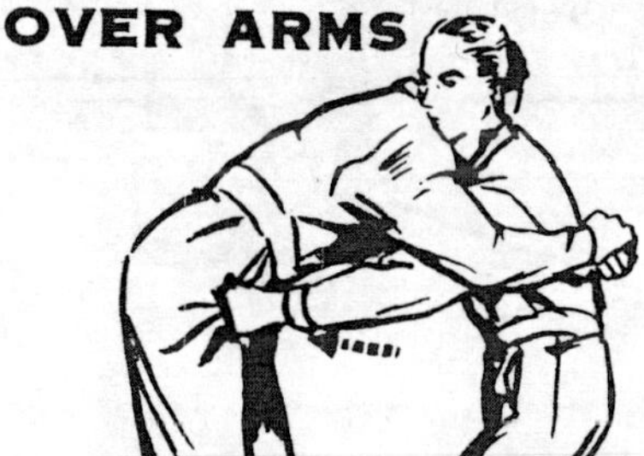

57

58

Drive thumbs or fist into opponent's groin forcing his hips back and body to bend forward.

Drive the knee into the groin and follow through with kick.

COUNTERS TO BEAR HUG

UNDER ARMS

Countering any kind of Bear Hug should be employed by immediate toe-or-heel kicking to the shins. When the hands are free—use them in vital spots.

59

Place thumb at base of opponent's nose and force head back or use V or hand and strike under nose. Fingers in the eyes will have same effect.

60

Grasp throat high with thumb and forefinger. Squeeze (dangerous). Coughing and choking results in the slightest pressure. Follow thru with trip to ground and complete pressure on ground.

61

Grasp shirt. Place the first two fingers of right hand quickly in hollow space, below Adam's Apple. Force two fingers down. Opponent will sit down. Follow thru with kick.

COUNTERS TO CHOKES

Use fingers immediately to the eyes. Temporary blinding is very effective. As soon as the eyes are touched your opponent will drop his hands – follow through immediately with blows or kicks.

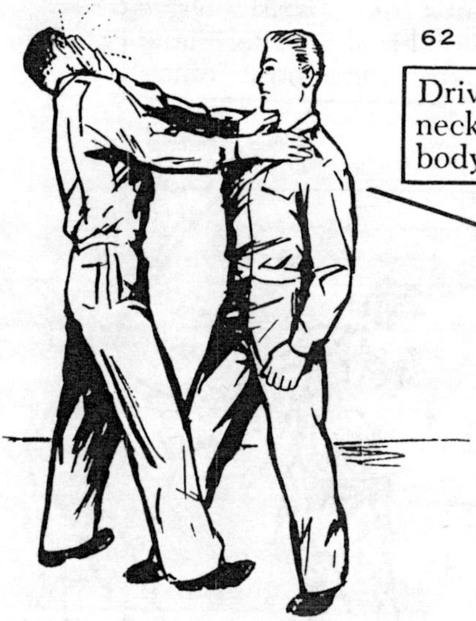

62

Drive edge of hand to neck using power of the body.

63

Reach through opponent's arms and grasp hair or ears.

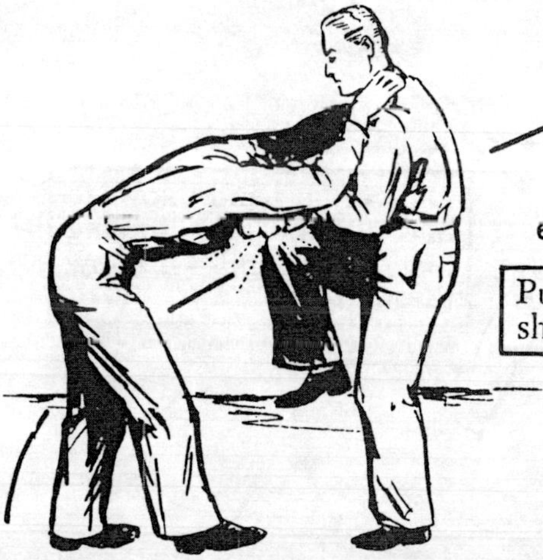

64

Pull head down to meet sharply raised knee.

COUNTERS TO CHOKES

65

Reach over with right hand grasping opponent's right wrist. Drive the palm of your left hand against his elbow joint.

66

Keeping opponent's right hand anchored to your shoulder, apply pressure to his elbow joint.

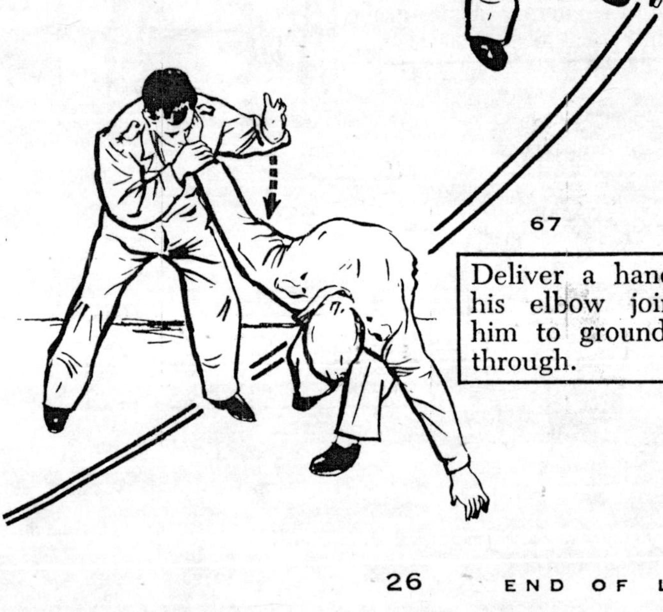

67

Deliver a hand blow to his elbow joint forcing him to ground. – Follow through.

KNEE KICK

A knee kick is easy to block. Pivot and block kick with hands, but grasp the leg and use his unbalance by lifting, pitching or throwing him to the ground. Be cautious in practice – it is dangerous.

68

Block right knee kick to the groin by pivoting to the right placing forearm on top of the knee.

69

Grasp collar with left hand; place right hand under raised right knee.

70

Follow thru by lifting right knee pushing opponent down and backward with head leading. (Dangerous.)

LEG KICK

71

Always block and grasp the foot when kicked. Opponent is off balance and can be thrown, pulled, or pushed into many effective falls. Maintain your own balance.

72

Place left leg in front of right leg, locking knees.

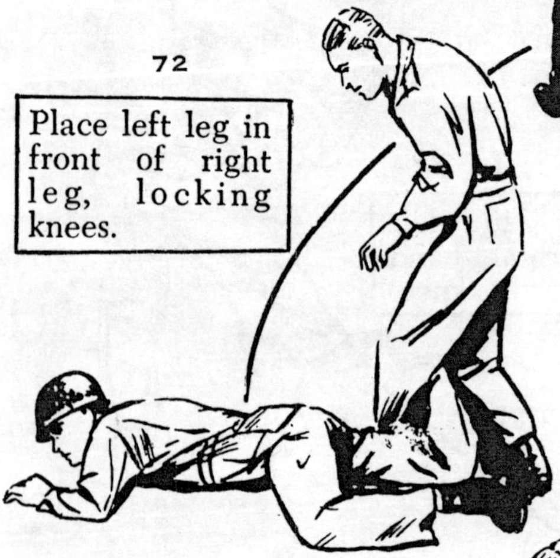

73

Twist or turn the foot to the right. Step forward and push opponent to the ground – face downward. A knee, crotch or straight push can be used. A quick kick will bring effective results.

74

Bend left leg–place the left knee and leg parallel to opponent's shoulders. Lean forward, lock leg and apply body pressure on leg. The hands should follow thru for a smudge, choke or helmet blow.

GROUND COUNTERS

When a man is thrown or jumped on, be sure he is locked. Do not let him roll or permit him to use his arms. Secure a punishing hold immediately — and smash thru. Don't let him see your position — lock him — pressure him — disable him.

Always lock opponent's arms. Spread and do not permit opponent to roll. Grasp right wrist and pull thru for hammerlock.

75

76

Repeat — Lock both hands and arms for double hammerlock.

Repeat with both hands.

77

78

Apply deep finger pressure in the soft fleshy part of shoulder. Pain severe. Raise head and smash to ground — or smudge face by leaning forward.

TAKE DOWNS

All Take Downs must have an element of surprise and a specific technique. It must be planned, must be effective and coordinated. The first contact must be certain – hold on to what you grab. Be prepared for his counter by employing another new counter.

79

80

Contact of hands and foot must be simultaneous.

The hands grab the top of the shoulders and pull back and down. Kick opponent's right knee, bending knees. Bring opponent to the ground. – Follow through.

81

Place right hand around opponent's throat. Place left hand on left shoulder. Raise knee. Squeeze, pull, kick and bring opponent to the ground. Follow thru with pressure on throat.

82

Grasp both ears — thumbs on top. Kick right knee or toe to base of spine. Opponent will come down quickly.

83

Place right knee in back of opponent's back. Force ears downward. Head will bend forward. Kick will complete movement.

84

Right hand grasps the collar and pulls backward as the left fist is driven into the small of the back or kidney. — Follow thru on ground.

85

Grasp the hair and pull backward. Use leg kick to knee or spine. — Knee can also be used.

STRANGLES

Strangling is effective and dangerous. It is the most deadly surprise attack in Hand-to-Hand Combat. Pressure on the throat causes irritation and with force—strangulation. It should be practiced under supervision. The "STOP" signal must be observed at all times.

Trip the opponent to the ground, continue choke with more pressure on throat. Lock arms and legs. (Dangerous.)

86

Cross collar choke. Grasp left side of collar (high) with left hand. Right hand grasps the right shoulder lapel (low). Place flat edge of hand against the throat. Increase pressure by pulling right arm.

87

88

As the pressure is applied, the right knee straightens and the left knee bends forcing your right shoulder forward against opponent's head. After opponent is brought to the ground, continue with the pressure. DANGEROUS.

89

The left hand is placed in back of the opponent's head as the thumb side of the wrist is driven into opponent's throat above the "Adams Apple." The head is then locked around the neck.

The Strangle –
The left hand grasps your own right hand. The head is forced forward by your right shoulder as the right arm is pulled to the rear driving the thin portion of the wrist above the "Adams Apple." Use the head for additional pressure.

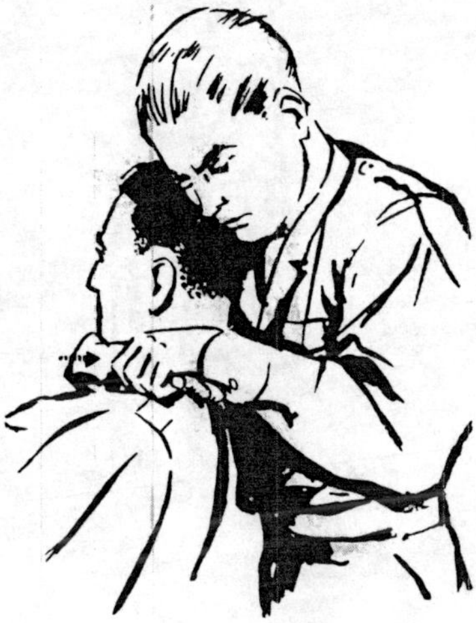

90

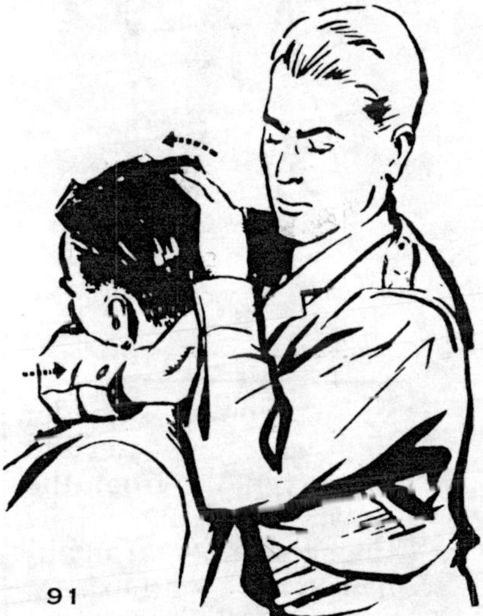

91

The right hand is then locked on your own left forearm or upper arm. The right arm applies pressure to the rear as the left hand forces the head forward.

END OF LESSON 4

COUNTERS TO BEAR HUG
UNDER ARMS

The use of the head, the sudden thrust of the elbow, the jab into the stomach, the twist and leg trip, and numerous other techniques should be employed in countering Bear Hugs. If your opponent is loose, take him to the ground with you, twisting so his back and head hit the ground – use him as a cushion.

92

Drive head backward. This will force him to break hold. Then follow thru with immediate trip.

93

Raise elbows shoulder high and pivot several times rapidly from the waist. Drive one elbow then the other to opponent's neck or face.

94

Opponent's head hidden behind shoulder.

95

Place your left hand on opponent's left knee as a base. Switch your left foot behind opponent's right foot.

96

Place hands under opponent's knees and strike the back of his right knee with your left knee.

97

Balance opponent over your left thigh and bring him to the ground by falling backward. Follow through with effective blows.

COUNTERS TO BEAR HUG
OVER ARMS

Elbows, arms, fingers are always vulnerable. The grab, must be sudden, the movement alert, and the follow through immediate. Any hesitation will give your opponent an opportunity to employ a counter.

98

Kick heel to instep, or toes, lower body by bending knees, and raise elbows shoulder high.

99

Place right foot outside of opponent's right foot. Lower body, grasp his right elbow with right hand holding it close to your chest. Left hand may grasp opponent's right wrist.

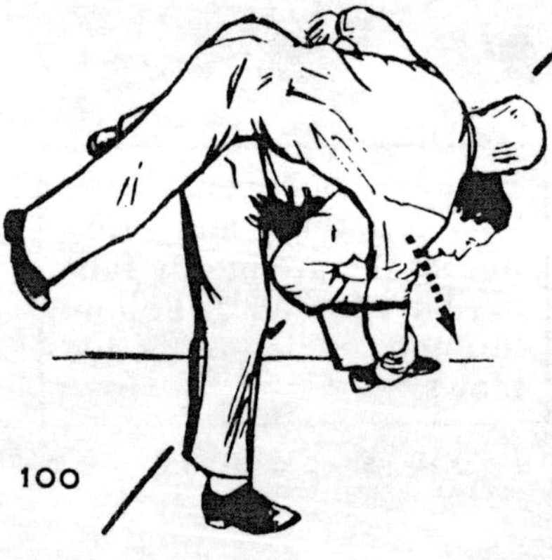
100

Bend swiftly from the waist throwing opponent over right shoulder. Hold arms and follow through with kick to head or neck.

COUNTERS TO FULL NELSON

105 Follow thru with a body slam to the ground.

COUNTERS TO STRANGLE

Try and beat the strangle. SENSE a person near you – Turn quickly and make every movement count. The pivot must be deliberate if that is your intention. Don't give away any feinting movements. The throws are good, but the techniques on the ground must be thorough.

Rear Strangle.

106

107

Place your right foot outside of opponent's right foot and lower your body by bending the knees. Turn head – place chin in bent arm.

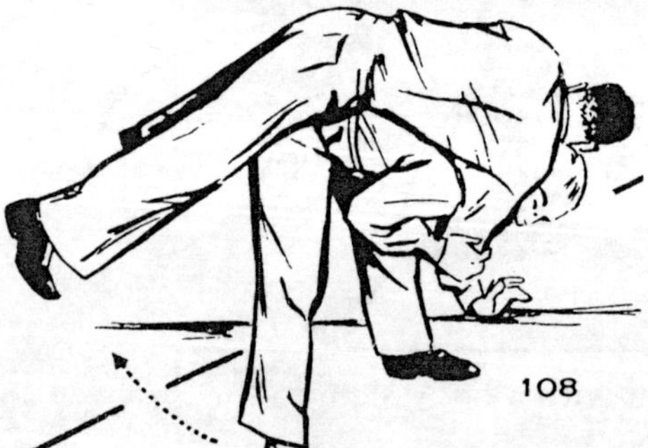

108

Bend swiftly from the waist throwing opponent over right shoulder. Hold arms and follow through with kick to head or neck.

Single Arm Strangle

109

110

Pivot the hips away. Step behind opponent leaving your left arm in front of his body.

Trip opponent's right knee with your left one as you strike backward with your left elbow driving him to the ground.

111

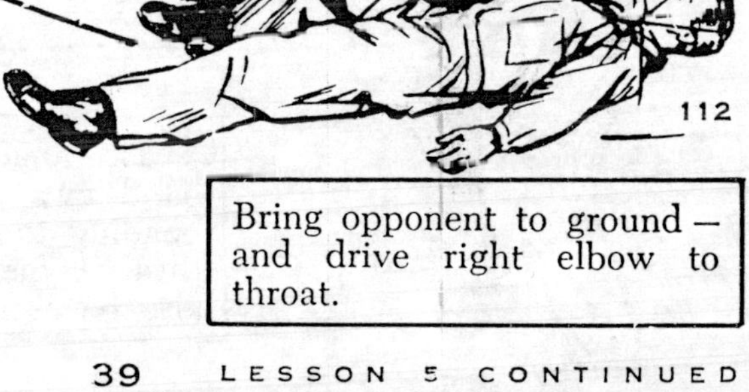

112

Bring opponent to ground — and drive right elbow to throat.

COUNTERS TO STRANGLE—

113

Double arm strangle. Grasp thumb or fingers of opponent with right hand. Grasp left elbow.

114

Twist fingers away from the body, causing arm to bend and relax. Place left hand on elbow locking arm.

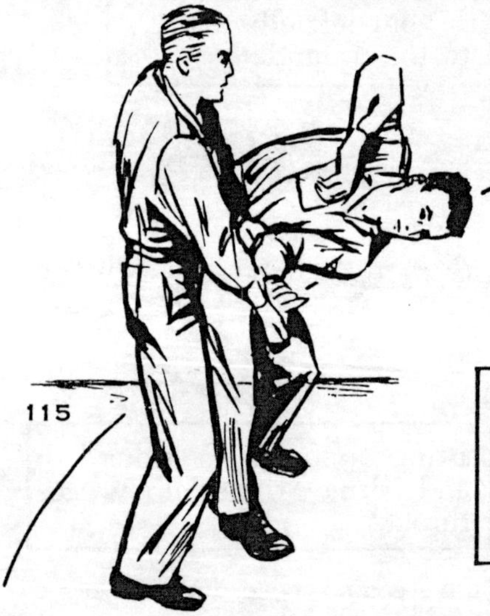

115

Twist fingers of either hand away, locking and bending the arm. Follow thru. A hammerlock can also be used.

LESSON 6 ADVANCED TECHNIQUES

SILENCING SENTRIES

116

Pull helmet quickly from the head. Place left hand on neck – get opponent off balance.

117

Smash helmet to head or back of neck.

118

Place both hands quickly on top and side of helmet. Lock arms of opponent with your own arms.

119

Force helmet forward and downward (exert sudden pressure). Opponent will bend on knees. Use shoulders and pressure on head on ground.

 LESSON 6 CONTINUED

SILENCING SENTRIES

120

Fix small loop of rope or wire. Have elbows parallel and low to block any movement of opponent's arms.

121

Place small loop over head and around neck. Elbows down.

122

Pull back and down quickly. (Dangerous.)

123

Apply knee pressure. Strangulation follows quickly. (Dangerous.)

The helmet which has been used as a hammer, bucket, and cooking utensil, is also a very dangerous weapon. It may protect a soldier from bullets, but also can be used on him as a bludgeon. There are many techniques that can be employed. It is an excellent counter against a trench tool or bayonet. If close to an enemy, it can be thrown at the face, as a temporary measure – then follow thru with ground tactics.

124

Reach over and grasp the leading edge of the helmet. The left forearm is placed across the base of the neck.

125

The fingers of the right hand grasp the helmet and start pulling backward. The fingers of the left hand grasps the right shoulder.

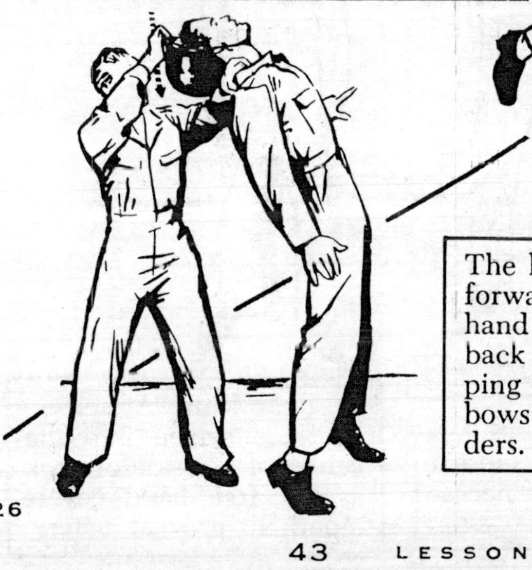

126

The left forearm pushes forward as the right hand pulls the helmet back and down, snapping the neck. Have elbows parallel to shoulders.

SILENCING SENTRIES

127

Right hand smudge and left blow to kidney – with knife.

128

Attack must be sudden and quiet. Jump if necessary.

129

Rear Knife Attack
Knife is driven into the center of the side of neck as the free hand covers mouth to prevent outcry.

COUNTERS TO KNIFE ATTACKS

Quick movement of arms and body is essential. Coordination – quick thinking and deliberate disarming must be made. Absolute confidence in your counter will surprise opponent as much as unbalance him. When he is disarmed, either use the same knife on him – or follow thru with effective blows.

130

Block the forearm.

131

Strike elbow joint with wrist of your free arm causing elbow to bend.

132

Lock free hand on forearm. Bend sharply at the waist. A trip will bring opponent to the ground. Follow thru.

LESSON 6 ADVANCED TECHNIQUES

COUNTERS TO KNIFE ATTACKS

133

Twist body to the left quickly and block knife with the forearm.

134

Grasp knife hand with the right hand, thumb on the back of the hand and the fingers around palm.

135

Either break wrist or plunge knife into opponent's body, by grasping his hand and locking fingers. – Follow through.

136

Straight thrust to stomach – twist trunk and block with forearm.

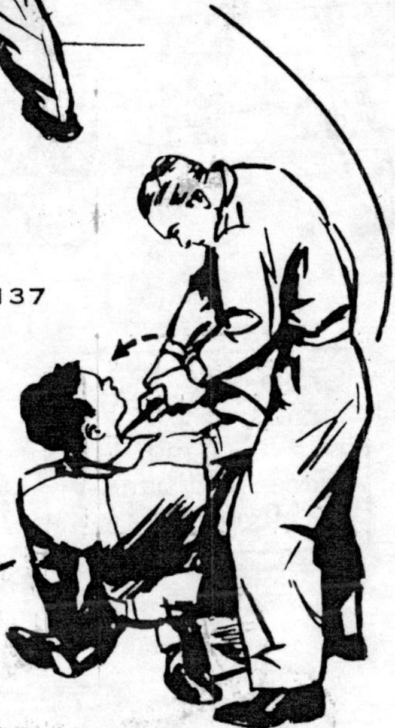

137

Lock opponent's fingers with the knife or bayonet – reinforcing with both hands. Bend wrist quickly, forcing opponent to the ground.

138

Place right knee on chest, and plunge knife or bayonet into the throat.

COUNTERS TO KNIFE ATTACKS

139

Block under-thrust – Twist or turn and block with fore-arm.

Grasp and lock fingers with right hand – place left hand on elbow.

140

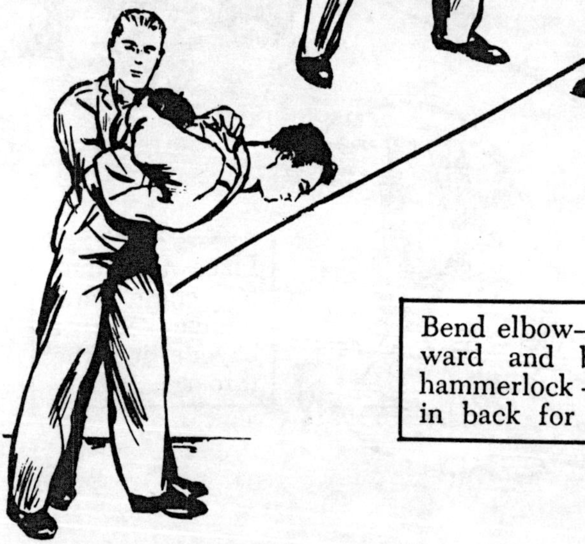

141

Bend elbow—bring downward and backward to hammerlock – force knife in back for follow thru.

PHYSICAL ACHIEVEMENT TESTS

PHYSICAL TESTING

The medical examination should be recognized as the most important test in any program designed to measure and develop physical and military fitness. Without a well organized and coordinated testing program, the possible weaknesses, deficiencies, and needs of men cannot be met satisfactorily. Tests stimulate interest and give men an opportunity to see where they stand in their all around ability with the rest of the group. Tests must be worth doing and must produce results that can be utilized for the improvement of the men and the training program.

The events should test strength, agility, endurance, co-ordination, speed, and adaptability. No classification system should be considered absolute; revisions should constantly be made as further records are tabulated. The standards should never be so high as to seem impossible of achievement nor so low as to discourage further interested effort. Strict supervision of the tests will have a great deal to do with the final results attained.

The test should be administered to each individual at periodic intervals throughout the training period in order to measure total progress from beginning to end. Look for improvement, interest, attitude and physical confidence.

"Physical confidence should be the main objective of the test."

HOW GOOD ARE YOU?

PA TEST NO. 1

CIRCLE YOUR SCORE

RATING	PULLUPS	PUSHUPS	SITUPS 3 MIN.	50 YD. CRAWL	50 YD. PIG-A-BACK	70 YD. AGILITY	350 YD. GRENADE	OBSTACLE COURSE	4 MILES
Superior	18 16	55 51	130 91	40 50	9	19	65		
Excellent	15 13	50 41	90 76	51 60	10	20 21	66 69		
Very Satisfactory	12 9	40 36	75 66	61 75	11 12	22 24	70 73		
Satisfactory	8 6	35 25	65 60	76 84	13 14	25 26	74 77		
Unsatisfactory	5	24	59	85	15	27	78		

PA TEST NO. 2

RATING	PULLUPS	PUSHUPS	SITUPS 3 MIN.	50 YD. CRAWL	50 YD. PIG-A-BACK	70 YD. AGILITY	350 YD. GRENADE	CONDITIONING COURSE	NINE MILES
Superior	19 17	56	131 97	41	8	18	64		
Excellent	16 14	55 46	96 82	42 58	9	19 20	65 68		
Very Satisfactory	13 10	45 41	81 72	59 68	10 11	21 23	69 72		
Satisfactory	9 7	40 30	71 66	69 79	12 13	24 25	73 75		
Unsatisfactory	6	29	65	80	14	26	76		

GENERAL RULES

PULLUPS – Individual must pull his weight up to the bar – chin above the bar. Elbows must be straight and arms completely extended on the return.

AGILITY RUN – Start from crouch position. (Run 10 yds; Crawl 10 yds – Run 10 yds; Creep 10 yds – Run 10 yds; Hurdle 10 yds – Run 10 yds) Total 70 Yards.

350 YARD GRENADE – Course 50 Yards; place a grenade or block of wood every ten yards. Five grenades. Take one grenade at a time and return to box. Then run 50 yards to finish.

PUSHUPS – Entire body must be kept in straight line. Chest only touches ground.

SITUPS – Start lying on back, hands clasped behind head. Ankles held – given three minutes to complete. On sit up, elbow must touch opposite knee each time.

50 YARD PICK A BACK – Pair off according to weight. Carry on back 50 Yards. The man carrying the weight is given the time. Change partners for other time.

50 YARD CRAWL – Start on stomachs in front of 50 yard barbed wire (20 inches) area. Crawl according to (FM-21-75). Time will be recorded when body rolls in slit trench.